The Plant Files

Create Your Own Plant Care Reference for Every Plant You Grow

Property of:

____

ISBN: 978-1-968971-30-4

Ironstone
Tennessee

Disclaimer and Terms of Use:

The author and publisher have attempted to be as accurate and complete as possible in creating this planner.

Although all attempts have been made to verify the information provided in this publication, the author and publisher assume no responsibility for errors, omissions, or contrary interpretations of the subject matter in this publication. Any perceived slights of specific people, business, or organizations are unintentional.

This content is for personal use only and may not be copied or resold in any form without prior written permission from the author.

Print copies of this book are available for book signings, special events, and retail gift shop purchases. For direct ship information, of copies of 10 or more, please contact the author at ironstone_tennessee@yahoo.com.

The Plant Files
Create Your Own Plant Care Reference for Every Plant You Grow
ISBN: 978-1-968971-30-4

www.ironstone-tennessee.com

Table of Contents

How to Use The Plant Files

This workbook is for the gardener who grows a little of everything. Each section is dedicated to recording your research, knowledge and experience with growing a specific type of plant. This allows you to keep your notes organized and helps you find the information you need quickly.

With that said, you should use this workbook however it best suits you. If you don't grow herbs, use that space for additional ornamentals, edibles or weed notes. Some plants will fit more than one category. Pick the one that fits your purposes best. If you grow lavender as an ornamental, put it with the perennial flowers. If you grow it as a medicinal or flavoring, put it with the herbs.

If you decide that you need your workbook to be ring bound or spiral bound, I've left wide enough inner margins that you should be able to remove the spine and punch the necessary holes without cutting off important parts of the book.

If you want to add page tabs to help you find the right section more quickly, there should be enough room in the outer margins to accommodate that.

Important Data

These pages keep your growing conditions and goals in one spot, so you don't lose your focus.

Plant Diseases and Pests

Every part of the country has its' gardening challenges. Even indoor plants are sometimes susceptible to disease and pests. A little research beforehand will help you be prepared to deal with problems as they arise. And a little note-keeping will help you remember what worked, and what didn't, later on.

Month-By-Month Lists

These lists help you remember time-sensitive garden chores, harvest windows and bloom seasons in your garden. It's helpful to have these lists handy when you plan your garden schedule.

Plant Files

The blank rectangle on the top right of each plant page offers space to sketch the plant, add a pressed leaf or flower from the plant, or add a photo of the plant for future reference.

The plant pages in each section are customized to the information that is important to know for each type of plant. We don't worry about cross-pollinators for houseplants, but that's critical information for fruit trees. Likewise, harvest times and companion plants are important to know for garden vegetables, but not so much for weeds.

The notes section on each plant page gives you plenty of room to observe how your plants handle drought, unexpected frosts, heavy rainfall, pressure from deer, and other environmental effects. You can note those plants that you grow for sentimental reasons (grandma grew these roses by the front porch), which apples make the best pie, or that your patch of walking onions came from your great-grandfather's homestead.

The Plant Files isn't meant to replace a good garden planner or journal, and it won't take the place of a good gardening manual either. Rather, it's a companion to those tools. A place to record information as you research the plants you want to grow, to record the characteristics, problems and benefits you observe in the plants you are already growing, and a reference to share with other gardeners. As you put it to work, it will become a unique reference, perfectly tailored to your property, the plants that you grow, and your style of gardening.

I wish you a very successful gardening journey.

Mel at Ironstone Tennessee

Important Dates and Data

Growing Zone	Soil Type(s)	Soil pH	Avg 1st Frost	Avg. Last Frost	Avg. Annual Rainfall

My Big Garden Goals:

My Smaller Garden Goals:

1.

2.

3.

4.

My Home Environment

Summer Temperature	Summer Humidity	Winter Temperature	Winter Humidity	Microclimates	Window Directions

My Best Houseplant Tips:

Fertilizers & Soil Amendments	Purpose	Potting Soils:

Plant Diseases

Disease / Organism	Plants Affected	Symptoms	Treatment	Prevention

Plant Pests

Pest	Plant Affected	Damage/ Symptoms	Treatment/ Prevention	Active Season

NOTES:

Notes:

Bloom Times

Bloom Times

January

February

March

April

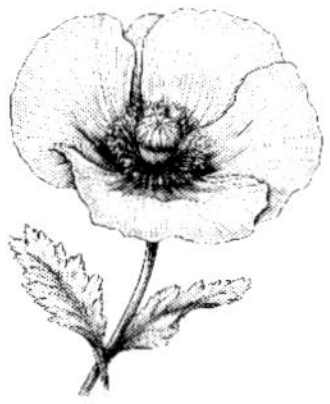

Bloom Times

May

June

July

August

Bloom Times

September

October

November

December

Harvest Times

Harvest Times

January

February

March

April

HARVEST TIMES

May

June

July

August

Harvest Times

September

October

November

December

Planting Times

Planting Times

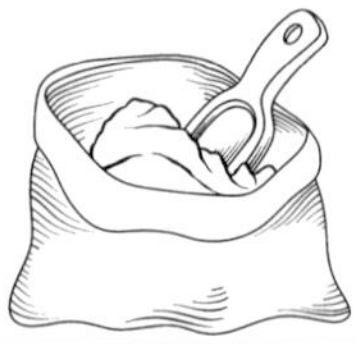

January

February

March

April

Planting Times

May

June

July

August

Planting Times

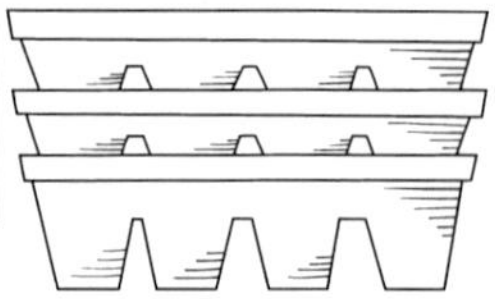

September

October

November

December

Chores by Month

Chores by Month

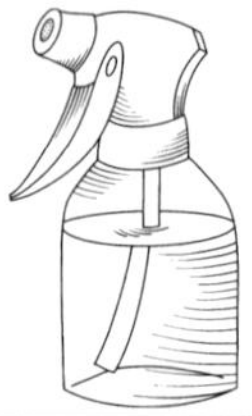

January

February

March

April

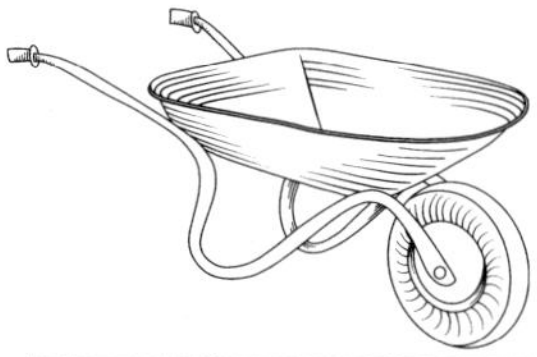

Chores by Month

May

June

July

August

Chores by Month

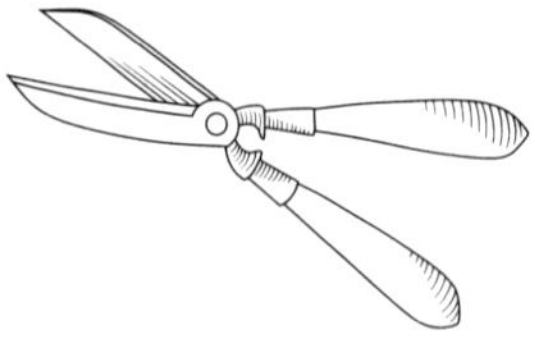

September

October

November

December

Notes:

Annual Fruits & Vegetables

Common Name		
Latin Name		
Cultivar		
Type		
Harvest Time		
Propagation		
Source		
Light Preference		
Soil Type		
Water Requirement		

Companion Plants	
Soil Amendments	
Common Diseases	
Pest Pressure	
Uses/Why I Grow It	

Notes:

Common Name		
Latin Name		
Cultivar		
Type		
Harvest Time		
Propagation		
Source		
Light Preference		
Soil Type		
Water Requirement		

Companion Plants	
Soil Amendments	
Common Diseases	
Pest Pressure	
Uses/Why I Grow It	

Notes:

Common Name		
Latin Name		
Cultivar		
Type		
Harvest Time		
Propagation		
Source		
Light Preference		
Soil Type		
Water Requirement		

Companion Plants	
Soil Amendments	
Common Diseases	
Pest Pressure	
Uses/Why I Grow It	

Notes:

Common Name		
Latin Name		
Cultivar		
Type		
Harvest Time		
Propagation		
Source		
Light Preference		
Soil Type		
Water Requirement		

Companion Plants	
Soil Amendments	
Common Diseases	
Pest Pressure	
Uses/Why I Grow It	

Notes:

Common Name	
Latin Name	
Cultivar	
Type	
Harvest Time	
Propagation	
Source	
Light Preference	
Soil Type	
Water Requirement	
Companion Plants	
Soil Amendments	
Common Diseases	
Pest Pressure	
Uses/Why I Grow It	

Notes:

Common Name	
Latin Name	
Cultivar	
Type	
Harvest Time	
Propagation	
Source	
Light Preference	
Soil Type	
Water Requirement	
Companion Plants	
Soil Amendments	
Common Diseases	
Pest Pressure	
Uses/Why I Grow It	

Notes:

Common Name		
Latin Name		
Cultivar		
Type		
Harvest Time		
Propagation		
Source		
Light Preference		
Soil Type		
Water Requirement		

Companion Plants	
Soil Amendments	
Common Diseases	
Pest Pressure	
Uses/Why I Grow It	

Notes:

Common Name		
Latin Name		
Cultivar		
Type		
Harvest Time		
Propagation		
Source		
Light Preference		
Soil Type		
Water Requirement		

Companion Plants	
Soil Amendments	
Common Diseases	
Pest Pressure	
Uses/Why I Grow It	

Notes:

Common Name	
Latin Name	
Cultivar	
Type	
Harvest Time	
Propagation	
Source	
Light Preference	
Soil Type	
Water Requirement	
Companion Plants	
Soil Amendments	
Common Diseases	
Pest Pressure	
Uses/Why I Grow It	

Notes:

Common Name	
Latin Name	
Cultivar	
Type	
Harvest Time	
Propagation	
Source	
Light Preference	
Soil Type	
Water Requirement	

Companion Plants	
Soil Amendments	
Common Diseases	
Pest Pressure	
Uses/Why I Grow It	

Notes:

Common Name		
Latin Name		
Cultivar		
Type		
Harvest Time		
Propagation		
Source		
Light Preference		
Soil Type		
Water Requirement		

Companion Plants	
Soil Amendments	
Common Diseases	
Pest Pressure	
Uses/Why I Grow It	

Notes:

Common Name		
Latin Name		
Cultivar		
Type		
Harvest Time		
Propagation		
Source		
Light Preference		
Soil Type		
Water Requirement		

Companion Plants	
Soil Amendments	
Common Diseases	
Pest Pressure	
Uses/Why I Grow It	

Notes:

Common Name		
Latin Name		
Cultivar		
Type		
Harvest Time		
Propagation		
Source		
Light Preference		
Soil Type		
Water Requirement		

Companion Plants	
Soil Amendments	
Common Diseases	
Pest Pressure	
Uses/Why I Grow It	

Notes:

Common Name		
Latin Name		
Cultivar		
Type		
Harvest Time		
Propagation		
Source		
Light Preference		
Soil Type		
Water Requirement		
Companion Plants		
Soil Amendments		
Common Diseases		
Pest Pressure		
Uses/Why I Grow It		

Notes:

Common Name		
Latin Name		
Cultivar		
Type		
Harvest Time		
Propagation		
Source		
Light Preference		
Soil Type		
Water Requirement		

Companion Plants	
Soil Amendments	
Common Diseases	
Pest Pressure	
Uses/Why I Grow It	

Notes:

Common Name		
Latin Name		
Cultivar		
Type		
Harvest Time		
Propagation		
Source		
Light Preference		
Soil Type		
Water Requirement		

Companion Plants	
Soil Amendments	
Common Diseases	
Pest Pressure	
Uses/Why I Grow It	

Notes:

Common Name		
Latin Name		
Cultivar		
Type		
Harvest Time		
Propagation		
Source		
Light Preference		
Soil Type		
Water Requirement		

Companion Plants	
Soil Amendments	
Common Diseases	
Pest Pressure	
Uses/Why I Grow It	

Notes:

Common Name	
Latin Name	
Cultivar	
Type	
Harvest Time	
Propagation	
Source	
Light Preference	
Soil Type	
Water Requirement	
Companion Plants	
Soil Amendments	
Common Diseases	
Pest Pressure	
Uses/Why I Grow It	

Notes:

Common Name		
Latin Name		
Cultivar		
Type		
Harvest Time		
Propagation		
Source		
Light Preference		
Soil Type		
Water Requirement		

Companion Plants	
Soil Amendments	
Common Diseases	
Pest Pressure	
Uses/Why I Grow It	

Notes:

Common Name	
Latin Name	
Cultivar	
Type	
Harvest Time	
Propagation	
Source	
Light Preference	
Soil Type	
Water Requirement	
Companion Plants	
Soil Amendments	
Common Diseases	
Pest Pressure	
Uses/Why I Grow It	

Notes:

Common Name	
Latin Name	
Cultivar	
Type	
Harvest Time	
Propagation	
Source	
Light Preference	
Soil Type	
Water Requirement	
Companion Plants	
Soil Amendments	
Common Diseases	
Pest Pressure	
Uses/Why I Grow It	

Notes:

NOTES:

Fruit & Nut Trees

Size/Root Stock		
Fruit Qualities		
Bloom Time		
Cross-Pollinator		
Harvest Time		
Source		
Disease Resistance		
Pruning		

Soil Amendments	
Diseases & Treatments	
Pests & Treatments	

Notes:

Size/Root Stock		
Fruit Qualities		
Bloom Time		
Cross-Pollinator		
Harvest Time		
Source		
Disease Resistance		
Pruning		

Soil Amendments	
Diseases & Treatments	
Pests & Treatments	

Notes:

Size/Root Stock		
Fruit Qualities		
Bloom Time		
Cross-Pollinator		
Harvest Time		
Source		
Disease Resistance		
Pruning		

Soil Amendments	
Diseases & Treatments	
Pests & Treatments	

Notes:

Size/Root Stock		
Fruit Qualities		
Bloom Time		
Cross-Pollinator		
Harvest Time		
Source		
Disease Resistance		
Pruning		

Soil Amendments	
Diseases & Treatments	
Pests & Treatments	

Notes:

Size/Root Stock		
Fruit Qualities		
Bloom Time		
Cross-Pollinator		
Harvest Time		
Source		
Disease Resistance		
Pruning		

Soil Amendments	
Diseases & Treatments	
Pests & Treatments	

Notes:

Size/Root Stock		
Fruit Qualities		
Bloom Time		
Cross-Pollinator		
Harvest Time		
Source		
Disease Resistance		
Pruning		

Soil Amendments	
Diseases & Treatments	
Pests & Treatments	

Notes:

Size/Root Stock		
Fruit Qualities		
Bloom Time		
Cross-Pollinator		
Harvest Time		
Source		
Disease Resistance		
Pruning		

Soil Amendments	
Diseases & Treatments	
Pests & Treatments	

Notes:

Size/Root Stock		
Fruit Qualities		
Bloom Time		
Cross-Pollinator		
Harvest Time		
Source		
Disease Resistance		
Pruning		

Soil Amendments	
Diseases & Treatments	
Pests & Treatments	

Notes:

Size/Root Stock		
Fruit Qualities		
Bloom Time		
Cross-Pollinator		
Harvest Time		
Source		
Disease Resistance		
Pruning		

Soil Amendments	
Diseases & Treatments	
Pests & Treatments	

Notes:

Size/Root Stock		
Fruit Qualities		
Bloom Time		
Cross-Pollinator		
Harvest Time		
Source		
Disease Resistance		
Pruning		

Soil Amendments	
Diseases & Treatments	
Pests & Treatments	

Notes:

Size/Root Stock		
Fruit Qualities		
Bloom Time		
Cross-Pollinator		
Harvest Time		
Source		
Disease Resistance		
Pruning		

Soil Amendments	
Diseases & Treatments	
Pests & Treatments	

Notes:

Size/Root Stock		
Fruit Qualities		
Bloom Time		
Cross-Pollinator		
Harvest Time		
Source		
Disease Resistance		
Pruning		

Soil Amendments	
Diseases & Treatments	
Pests & Treatments	

Notes:

NOTES:

Annual Flowers

Common Name	
Latin Name	
Cultivar	
Type	
Bloom Time	
Fragrance	
Source	
Light Preference	
Soil Type	
Water Requirement	
Bloom Characteristics	
Soil Amendments	
Diseases & Control	
Pests & Control	

Notes:

Common Name	
Latin Name	
Cultivar	
Type	
Bloom Time	
Fragrance	
Source	
Light Preference	
Soil Type	
Water Requirement	
Bloom Characteristics	
Soil Amendments	
Diseases & Control	
Pests & Control	

Notes:

Common Name	
Latin Name	
Cultivar	
Type	
Bloom Time	
Fragrance	
Source	
Light Preference	
Soil Type	
Water Requirement	
Bloom Characteristics	
Soil Amendments	
Diseases & Control	
Pests & Control	

Notes:

Common Name	
Latin Name	
Cultivar	
Type	
Bloom Time	
Fragrance	
Source	
Light Preference	
Soil Type	
Water Requirement	
Bloom Characteristics	
Soil Amendments	
Diseases & Control	
Pests & Control	

Notes:

Common Name	
Latin Name	
Cultivar	
Type	
Bloom Time	
Fragrance	
Source	
Light Preference	
Soil Type	
Water Requirement	
Bloom Characteristics	
Soil Amendments	
Diseases & Control	
Pests & Control	

Notes:

Common Name	
Latin Name	
Cultivar	
Type	
Bloom Time	
Fragrance	
Source	
Light Preference	
Soil Type	
Water Requirement	
Bloom Characteristics	
Soil Amendments	
Diseases & Control	
Pests & Control	

Notes:

Common Name	
Latin Name	
Cultivar	
Type	
Bloom Time	
Fragrance	
Source	
Light Preference	
Soil Type	
Water Requirement	
Bloom Characteristics	
Soil Amendments	
Diseases & Control	
Pests & Control	

Notes:

Common Name	
Latin Name	
Cultivar	
Type	
Bloom Time	
Fragrance	
Source	
Light Preference	
Soil Type	
Water Requirement	
Bloom Characteristics	
Soil Amendments	
Diseases & Control	
Pests & Control	

Notes:

Common Name	
Latin Name	
Cultivar	
Type	
Bloom Time	
Fragrance	
Source	
Light Preference	
Soil Type	
Water Requirement	
Bloom Characteristics	
Soil Amendments	
Diseases & Control	
Pests & Control	

Notes:

Common Name	
Latin Name	
Cultivar	
Type	
Bloom Time	
Fragrance	
Source	
Light Preference	
Soil Type	
Water Requirement	
Bloom Characteristics	
Soil Amendments	
Diseases & Control	
Pests & Control	

Notes:

Common Name	
Latin Name	
Cultivar	
Type	
Bloom Time	
Fragrance	
Source	
Light Preference	
Soil Type	
Water Requirement	
Bloom Characteristics	
Soil Amendments	
Diseases & Control	
Pests & Control	

Notes:

Common Name	
Latin Name	
Cultivar	
Type	
Bloom Time	
Fragrance	
Source	
Light Preference	
Soil Type	
Water Requirement	
Bloom Characteristics	
Soil Amendments	
Diseases & Control	
Pests & Control	

Notes:

Notes:

PERENNIAL FLOWERS

Common Name		
Latin Name		
Cultivar		
Type		
Bloom Time		
Fragrance		
Source		
Light Preference		
Soil Type		
Water Requirement		
Bloom Characteristics		
Soil Amendments		
Diseases & Control		
Pests & Control		

Notes:

Common Name	
Latin Name	
Cultivar	
Type	
Bloom Time	
Fragrance	
Source	
Light Preference	
Soil Type	
Water Requirement	
Bloom Characteristics	
Soil Amendments	
Diseases & Control	
Pests & Control	

Notes:

Common Name	
Latin Name	
Cultivar	
Type	
Bloom Time	
Fragrance	
Source	
Light Preference	
Soil Type	
Water Requirement	
Bloom Characteristics	
Soil Amendments	
Diseases & Control	
Pests & Control	

Notes:

Common Name		
Latin Name		
Cultivar		
Type		
Bloom Time		
Fragrance		
Source		
Light Preference		
Soil Type		
Water Requirement		
Bloom Characteristics		
Soil Amendments		
Diseases & Control		
Pests & Control		

Notes:

Common Name	
Latin Name	
Cultivar	
Type	
Bloom Time	
Fragrance	
Source	
Light Preference	
Soil Type	
Water Requirement	
Bloom Characteristics	
Soil Amendments	
Diseases & Control	
Pests & Control	

Notes:

Common Name	
Latin Name	
Cultivar	
Type	
Bloom Time	
Fragrance	
Source	
Light Preference	
Soil Type	
Water Requirement	
Bloom Characteristics	
Soil Amendments	
Diseases & Control	
Pests & Control	

Notes:

Common Name	
Latin Name	
Cultivar	
Type	
Bloom Time	
Fragrance	
Source	
Light Preference	
Soil Type	
Water Requirement	
Bloom Characteristics	
Soil Amendments	
Diseases & Control	
Pests & Control	

Notes:

Common Name	
Latin Name	
Cultivar	
Type	
Bloom Time	
Fragrance	
Source	
Light Preference	
Soil Type	
Water Requirement	
Bloom Characteristics	
Soil Amendments	
Diseases & Control	
Pests & Control	

Notes:

Common Name	
Latin Name	
Cultivar	
Type	
Bloom Time	
Fragrance	
Source	
Light Preference	
Soil Type	
Water Requirement	
Bloom Characteristics	
Soil Amendments	
Diseases & Control	
Pests & Control	

Notes:

Common Name	
Latin Name	
Cultivar	
Type	
Bloom Time	
Fragrance	
Source	
Light Preference	
Soil Type	
Water Requirement	
Bloom Characteristics	
Soil Amendments	
Diseases & Control	
Pests & Control	

Notes:

Common Name	
Latin Name	
Cultivar	
Type	
Bloom Time	
Fragrance	
Source	
Light Preference	
Soil Type	
Water Requirement	
Bloom Characteristics	
Soil Amendments	
Diseases & Control	
Pests & Control	

Notes:

Common Name	
Latin Name	
Cultivar	
Type	
Bloom Time	
Fragrance	
Source	
Light Preference	
Soil Type	
Water Requirement	
Bloom Characteristics	
Soil Amendments	
Diseases & Control	
Pests & Control	

Notes:

Common Name	
Latin Name	
Cultivar	
Type	
Bloom Time	
Fragrance	
Source	
Light Preference	
Soil Type	
Water Requirement	
Bloom Characteristics	
Soil Amendments	
Diseases & Control	
Pests & Control	

Notes:

Common Name	
Latin Name	
Cultivar	
Type	
Bloom Time	
Fragrance	
Source	
Light Preference	
Soil Type	
Water Requirement	
Bloom Characteristics	
Soil Amendments	
Diseases & Control	
Pests & Control	

Notes:

Common Name	
Latin Name	
Cultivar	
Type	
Bloom Time	
Fragrance	
Source	
Light Preference	
Soil Type	
Water Requirement	
Bloom Characteristics	
Soil Amendments	
Diseases & Control	
Pests & Control	

Notes:

Common Name	
Latin Name	
Cultivar	
Type	
Bloom Time	
Fragrance	
Source	
Light Preference	
Soil Type	
Water Requirement	
Bloom Characteristics	
Soil Amendments	
Diseases & Control	
Pests & Control	

Notes:

Common Name	
Latin Name	
Cultivar	
Type	
Bloom Time	
Fragrance	
Source	
Light Preference	
Soil Type	
Water Requirement	
Bloom Characteristics	
Soil Amendments	
Diseases & Control	
Pests & Control	

Notes:

Common Name	
Latin Name	
Cultivar	
Type	
Bloom Time	
Fragrance	
Source	
Light Preference	
Soil Type	
Water Requirement	
Bloom Characteristics	
Soil Amendments	
Diseases & Control	
Pests & Control	

Notes:

Common Name	
Latin Name	
Cultivar	
Type	
Bloom Time	
Fragrance	
Source	
Light Preference	
Soil Type	
Water Requirement	
Bloom Characteristics	
Soil Amendments	
Diseases & Control	
Pests & Control	

Notes:

Common Name		
Latin Name		
Cultivar		
Type		
Bloom Time		
Fragrance		
Source		
Light Preference		
Soil Type		
Water Requirement		
Bloom Characteristics		
Soil Amendments		
Diseases & Control		
Pests & Control		

Notes:

Common Name	
Latin Name	
Cultivar	
Type	
Bloom Time	
Fragrance	
Source	
Light Preference	
Soil Type	
Water Requirement	
Bloom Characteristics	
Soil Amendments	
Diseases & Control	
Pests & Control	

Notes:

Common Name	
Latin Name	
Cultivar	
Type	
Bloom Time	
Fragrance	
Source	
Light Preference	
Soil Type	
Water Requirement	
Bloom Characteristics	
Soil Amendments	
Diseases & Control	
Pests & Control	

Notes:

Notes:

Ornamental Trees & Shrubs

Common Name	
Latin Name	
Cultivar	
Type	
Bloom Time	
Fragrance	
Source	
Light Preference	
Soil Type	
Water Requirement	
Ornamental Characteristics	
Soil Amendments	
Diseases & Control	
Pests & Control	

Notes:

Common Name	
Latin Name	
Cultivar	
Type	
Bloom Time	
Fragrance	
Source	
Light Preference	
Soil Type	
Water Requirement	
Ornamental Characteristics	
Soil Amendments	
Diseases & Control	
Pests & Control	

Notes:

Common Name	
Latin Name	
Cultivar	
Type	
Bloom Time	
Fragrance	
Source	
Light Preference	
Soil Type	
Water Requirement	
Ornamental Characteristics	
Soil Amendments	
Diseases & Control	
Pests & Control	

Notes:

Common Name	
Latin Name	
Cultivar	
Type	
Bloom Time	
Fragrance	
Source	
Light Preference	
Soil Type	
Water Requirement	
Ornamental Characteristics	
Soil Amendments	
Diseases & Control	
Pests & Control	

Notes:

Common Name	
Latin Name	
Cultivar	
Type	
Bloom Time	
Fragrance	
Source	
Light Preference	
Soil Type	
Water Requirement	
Ornamental Characteristics	
Soil Amendments	
Diseases & Control	
Pests & Control	

Notes:

Common Name	
Latin Name	
Cultivar	
Type	
Bloom Time	
Fragrance	
Source	
Light Preference	
Soil Type	
Water Requirement	
Ornamental Characteristics	
Soil Amendments	
Diseases & Control	
Pests & Control	

Notes:

Common Name	
Latin Name	
Cultivar	
Type	
Bloom Time	
Fragrance	
Source	
Light Preference	
Soil Type	
Water Requirement	
Ornamental Characteristics	
Soil Amendments	
Diseases & Control	
Pests & Control	

Notes:

Common Name	
Latin Name	
Cultivar	
Type	
Bloom Time	
Fragrance	
Source	
Light Preference	
Soil Type	
Water Requirement	
Ornamental Characteristics	
Soil Amendments	
Diseases & Control	
Pests & Control	

Notes:

Common Name	
Latin Name	
Cultivar	
Type	
Bloom Time	
Fragrance	
Source	
Light Preference	
Soil Type	
Water Requirement	
Ornamental Characteristics	
Soil Amendments	
Diseases & Control	
Pests & Control	

Notes:

Common Name	
Latin Name	
Cultivar	
Type	
Bloom Time	
Fragrance	
Source	
Light Preference	
Soil Type	
Water Requirement	
Ornamental Characteristics	
Soil Amendments	
Diseases & Control	
Pests & Control	

Notes:

Common Name	
Latin Name	
Cultivar	
Type	
Bloom Time	
Fragrance	
Source	
Light Preference	
Soil Type	
Water Requirement	
Ornamental Characteristics	
Soil Amendments	
Diseases & Control	
Pests & Control	

Notes:

Common Name	
Latin Name	
Cultivar	
Type	
Bloom Time	
Fragrance	
Source	
Light Preference	
Soil Type	
Water Requirement	
Ornamental Characteristics	
Soil Amendments	
Diseases & Control	
Pests & Control	

Notes:

Notes:

Herbs

Common Name		
Latin Name		
Harvest Time		
Planting & Propagation		
Source		
Light Preference		
Soil Type		
Water Requirement		
Companion Plants		
Soil Amendments		

Cautions:

Uses:

Notes:

Common Name	
Latin Name	
Harvest Time	
Planting & Propagation	
Source	
Light Preference	
Soil Type	
Water Requirement	
Companion Plants	
Soil Amendments	

***Cautions*:**

Uses:

Notes:

Common Name		
Latin Name		
Harvest Time		
Planting & Propagation		
Source		
Light Preference		
Soil Type		
Water Requirement		
Companion Plants		
Soil Amendments		

Cautions:

Uses:

Notes:

Common Name		
Latin Name		
Harvest Time		
Planting & Propagation		
Source		
Light Preference		
Soil Type		
Water Requirement		
Companion Plants		
Soil Amendments		

Cautions:

Uses:

Notes:

Common Name		
Latin Name		
Harvest Time		
Planting & Propagation		
Source		
Light Preference		
Soil Type		
Water Requirement		
Companion Plants		
Soil Amendments		

Cautions:

Uses:

Notes:

Common Name		
Latin Name		
Harvest Time		
Planting & Propagation		
Source		
Light Preference		
Soil Type		
Water Requirement		
Companion Plants		
Soil Amendments		

Cautions:

Uses:

Notes:

Common Name	
Latin Name	
Harvest Time	
Planting & Propagation	
Source	
Light Preference	
Soil Type	
Water Requirement	
Companion Plants	
Soil Amendments	

Cautions:

Uses:

Notes:

Common Name	
Latin Name	
Harvest Time	
Planting & Propagation	
Source	
Light Preference	
Soil Type	
Water Requirement	
Companion Plants	
Soil Amendments	

Cautions:

Uses:

Notes:

Common Name		
Latin Name		
Harvest Time		
Planting & Propagation		
Source		
Light Preference		
Soil Type		
Water Requirement		
Companion Plants		
Soil Amendments		

Cautions:

Uses:

Notes:

Common Name	
Latin Name	
Harvest Time	
Planting & Propagation	
Source	
Light Preference	
Soil Type	
Water Requirement	
Companion Plants	
Soil Amendments	

Cautions:

Uses:

Notes:

Common Name		
Latin Name		
Harvest Time		
Planting & Propagation		
Source		
Light Preference		
Soil Type		
Water Requirement		
Companion Plants		
Soil Amendments		

Cautions:

Uses:

Notes:

Common Name		
Latin Name		
Harvest Time		
Planting & Propagation		
Source		
Light Preference		
Soil Type		
Water Requirement		
Companion Plants		
Soil Amendments		

Cautions:

Uses:

Notes:

Notes:

Potted Plants

Common Name	
Latin Name & Cultivar	
Type	
Propagation	
Light Preference	
Exposure	
Humidity	
Temperature	
Soil Type	
Water Requirements	
Soil Amendments	
Growth Habit	
Leaf Characteristics	
Blooms	

Notes:

Common Name	
Latin Name & Cultivar	
Type	
Propagation	
Light Preference	
Exposure	
Humidity	
Temperature	
Soil Type	
Water Requirements	
Soil Amendments	
Growth Habit	
Leaf Characteristics	
Blooms	

Notes:

Common Name	
Latin Name & Cultivar	
Type	
Propagation	
Light Preference	
Exposure	
Humidity	
Temperature	
Soil Type	
Water Requirements	
Soil Amendments	
Growth Habit	
Leaf Characteristics	
Blooms	

Notes:

Common Name	
Latin Name & Cultivar	
Type	
Propagation	
Light Preference	
Exposure	
Humidity	
Temperature	
Soil Type	
Water Requirements	
Soil Amendments	
Growth Habit	
Leaf Characteristics	
Blooms	

Notes:

Common Name	
Latin Name & Cultivar	
Type	
Propagation	
Light Preference	
Exposure	
Humidity	
Temperature	
Soil Type	
Water Requirements	
Soil Amendments	
Growth Habit	
Leaf Characteristics	
Blooms	

Notes:

Common Name	
Latin Name & Cultivar	
Type	
Propagation	
Light Preference	
Exposure	
Humidity	
Temperature	
Soil Type	
Water Requirements	
Soil Amendments	
Growth Habit	
Leaf Characteristics	
Blooms	

Notes:

Common Name		
Latin Name & Cultivar		
Type		
Propagation		
Light Preference		
Exposure		
Humidity		
Temperature		
Soil Type		
Water Requirements		
Soil Amendments		
Growth Habit		
Leaf Characteristics		
Blooms		

Notes:

Common Name	
Latin Name & Cultivar	
Type	
Propagation	
Light Preference	
Exposure	
Humidity	
Temperature	
Soil Type	
Water Requirements	
Soil Amendments	
Growth Habit	
Leaf Characteristics	
Blooms	

Notes:

Common Name	
Latin Name & Cultivar	
Type	
Propagation	
Light Preference	
Exposure	
Humidity	
Temperature	
Soil Type	
Water Requirements	
Soil Amendments	
Growth Habit	
Leaf Characteristics	
Blooms	

Notes:

Common Name	
Latin Name & Cultivar	
Type	
Propagation	
Light Preference	
Exposure	
Humidity	
Temperature	
Soil Type	
Water Requirements	
Soil Amendments	
Growth Habit	
Leaf Characteristics	
Blooms	

Notes:

Common Name	
Latin Name & Cultivar	
Type	
Propagation	
Light Preference	
Exposure	
Humidity	
Temperature	
Soil Type	
Water Requirements	
Soil Amendments	
Growth Habit	
Leaf Characteristics	
Blooms	

Notes:

Common Name		
Latin Name & Cultivar		
Type		
Propagation		
Light Preference		
Exposure		
Humidity		
Temperature		
Soil Type		
Water Requirements		
Soil Amendments		
Growth Habit		
Leaf Characteristics		
Blooms		

Notes:

Notes:

Perennial Fruits & Vegetables

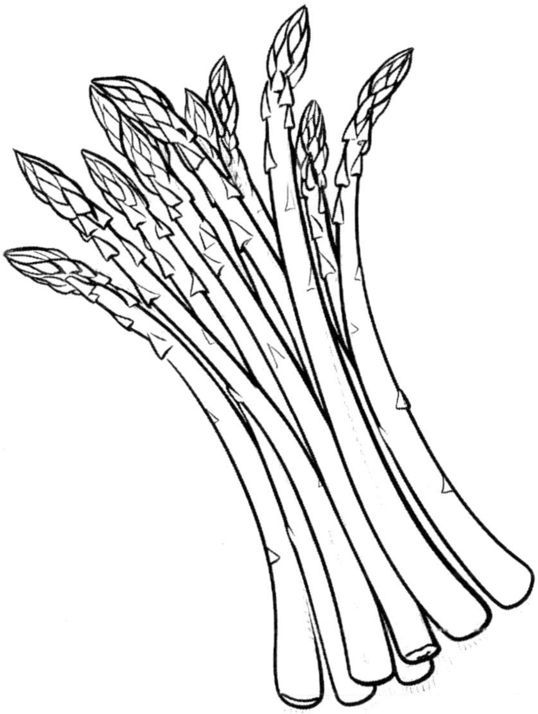

Common Name	
Latin Name	
Cultivar	
Type	
Harvest Time	
Propagation	
Source	
Light Preference	
Soil Type	
Water Requirement	

Companion Plants	
Soil Amendments	
Common Diseases	
Pest Pressure	
Uses/Why I Grow It	

Notes:

Common Name		
Latin Name		
Cultivar		
Type		
Harvest Time		
Propagation		
Source		
Light Preference		
Soil Type		
Water Requirement		

Companion Plants	
Soil Amendments	
Common Diseases	
Pest Pressure	
Uses/Why I Grow It	

Notes:

Common Name		
Latin Name		
Cultivar		
Type		
Harvest Time		
Propagation		
Source		
Light Preference		
Soil Type		
Water Requirement		

Companion Plants	
Soil Amendments	
Common Diseases	
Pest Pressure	
Uses/Why I Grow It	

Notes:

Common Name		
Latin Name		
Cultivar		
Type		
Harvest Time		
Propagation		
Source		
Light Preference		
Soil Type		
Water Requirement		

Companion Plants	
Soil Amendments	
Common Diseases	
Pest Pressure	
Uses/Why I Grow It	

Notes:

Common Name	
Latin Name	
Cultivar	
Type	
Harvest Time	
Propagation	
Source	
Light Preference	
Soil Type	
Water Requirement	

Companion Plants	
Soil Amendments	
Common Diseases	
Pest Pressure	
Uses/Why I Grow It	

Notes:

Common Name	
Latin Name	
Cultivar	
Type	
Harvest Time	
Propagation	
Source	
Light Preference	
Soil Type	
Water Requirement	

Companion Plants	
Soil Amendments	
Common Diseases	
Pest Pressure	
Uses/Why I Grow It	

Notes:

Common Name		
Latin Name		
Cultivar		
Type		
Harvest Time		
Propagation		
Source		
Light Preference		
Soil Type		
Water Requirement		

Companion Plants	
Soil Amendments	
Common Diseases	
Pest Pressure	
Uses/Why I Grow It	

Notes:

Common Name	
Latin Name	
Cultivar	
Type	
Harvest Time	
Propagation	
Source	
Light Preference	
Soil Type	
Water Requirement	
Companion Plants	
Soil Amendments	
Common Diseases	
Pest Pressure	
Uses/Why I Grow It	

Notes:

Common Name	
Latin Name	
Cultivar	
Type	
Harvest Time	
Propagation	
Source	
Light Preference	
Soil Type	
Water Requirement	

Companion Plants	
Soil Amendments	
Common Diseases	
Pest Pressure	
Uses/Why I Grow It	

Notes:

Common Name	
Latin Name	
Cultivar	
Type	
Harvest Time	
Propagation	
Source	
Light Preference	
Soil Type	
Water Requirement	

Companion Plants	
Soil Amendments	
Common Diseases	
Pest Pressure	
Uses/Why I Grow It	

Notes:

Notes:

Wild Plants

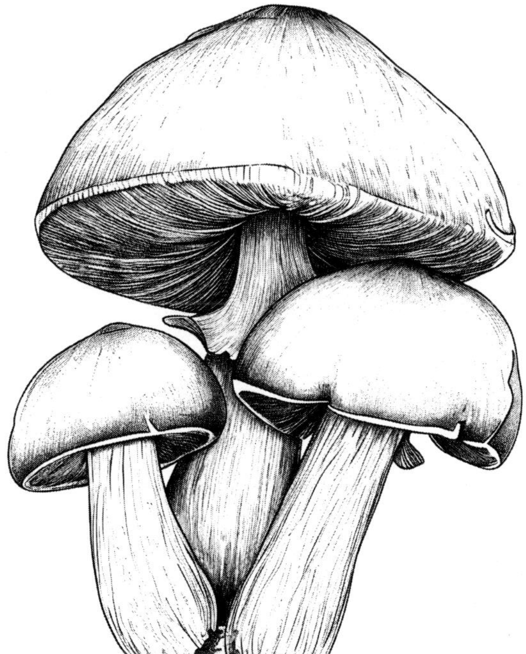

Plant Name	
Date	
Location(s)	
Companions	
Uses & Benefits	
When to Harvest	
Part(s) Used	

Aroma		Texture	
Color		Leaf Shape	
Warnings		Dangerous Look-alikes	

Notes:

Plant Name	
Date	
Location(s)	
Companions	
Uses & Benefits	
When to Harvest	
Part(s) Used	

Aroma		Texture	
Color		Leaf Shape	
Warnings		Dangerous Look-alikes	

Notes:

Plant Name	
Date	
Location(s)	
Companions	
Uses & Benefits	
When to Harvest	
Part(s) Used	

Aroma		Texture	
Color		Leaf Shape	
Warnings		Dangerous Look-alikes	

Notes:

Plant Name		
Date		
Location(s)		
Companions		
Uses & Benefits		
When to Harvest		
Part(s) Used		

Aroma		Texture	
Color		Leaf Shape	
Warnings		Dangerous Look-alikes	

Notes:

Plant Name		
Date		
Location(s)		
Companions		
Uses & Benefits		
When to Harvest		
Part(s) Used		

Aroma		Texture	
Color		Leaf Shape	
Warnings		Dangerous Look-alikes	

Notes:

Plant Name	
Date	
Location(s)	
Companions	
Uses & Benefits	
When to Harvest	
Part(s) Used	

Aroma		Texture	
Color		Leaf Shape	
Warnings		Dangerous Look-alikes	

Notes:

Plant Name		
Date		
Location(s)		
Companions		
Uses & Benefits		
When to Harvest		
Part(s) Used		

Aroma		Texture	
Color		Leaf Shape	
Warnings		Dangerous Look-alikes	

Notes:

Plant Name	
Date	
Location(s)	
Companions	
Uses & Benefits	
When to Harvest	
Part(s) Used	

Aroma		Texture	
Color		Leaf Shape	
Warnings		Dangerous Look-alikes	

Notes:

Plant Name		
Date		
Location(s)		
Companions		
Uses & Benefits		
When to Harvest		
Part(s) Used		

Aroma		Texture	
Color		Leaf Shape	
Warnings		Dangerous Look-alikes	

Notes:

Plant Name	
Date	
Location(s)	
Companions	
Uses & Benefits	
When to Harvest	
Part(s) Used	

Aroma		Texture	
Color		Leaf Shape	
Warnings		Dangerous Look-alikes	

Notes:

Plant Name	
Date	
Location(s)	
Companions	
Uses & Benefits	
When to Harvest	
Part(s) Used	

Aroma		Texture	
Color		Leaf Shape	
Warnings		Dangerous Look-alikes	

Notes:

Plant Name	
Date	
Location(s)	
Companions	
Uses & Benefits	
When to Harvest	
Part(s) Used	

Aroma		Texture	
Color		Leaf Shape	
Warnings		Dangerous Look-alikes	

Notes:

NOTES:

Weeds

Common Name	
Latin Name	
Origin	
Signals	
Soil Preference	
Benefits/Uses	
Appearance	
Control Method	

Common Name	
Latin Name	
Origin	
Signals	
Soil Preference	
Benefits/Uses	
Appearance	
Control Method	

Common Name	
Latin Name	
Origin	
Signals	
Soil Preference	
Benefits/Uses	
Appearance	
Control Method	

Common Name	
Latin Name	
Origin	
Signals	
Soil Preference	
Benefits/Uses	
Appearance	
Control Method	

Common Name	
Latin Name	
Origin	
Signals	
Soil Preference	
Benefits/Uses	
Appearance	
Control Method	

Common Name	
Latin Name	
Origin	
Signals	
Soil Preference	
Benefits/Uses	
Appearance	
Control Method	

Common Name		
Latin Name		
Origin		
Signals		
Soil Preference		
Benefits/Uses		
Appearance		
Control Method		

Common Name		
Latin Name		
Origin		
Signals		
Soil Preference		
Benefits/Uses		
Appearance		
Control Method		

Common Name		
Latin Name		
Origin		
Signals		
Soil Preference		
Benefits/Uses		
Appearance		
Control Method		

Common Name		
Latin Name		
Origin		
Signals		
Soil Preference		
Benefits/Uses		
Appearance		
Control Method		

Common Name	
Latin Name	
Origin	
Signals	
Soil Preference	
Benefits/Uses	
Appearance	
Control Method	

Common Name	
Latin Name	
Origin	
Signals	
Soil Preference	
Benefits/Uses	
Appearance	
Control Method	

Common Name		
Latin Name		
Origin		
Signals		
Soil Preference		
Benefits/Uses		
Appearance		
Control Method		

Common Name		
Latin Name		
Origin		
Signals		
Soil Preference		
Benefits/Uses		
Appearance		
Control Method		

Common Name		
Latin Name		
Origin		
Signals		
Soil Preference		
Benefits/Uses		
Appearance		
Control Method		

Common Name		
Latin Name		
Origin		
Signals		
Soil Preference		
Benefits/Uses		
Appearance		
Control Method		

Common Name		
Latin Name		
Origin		
Signals		
Soil Preference		
Benefits/Uses		
Appearance		
Control Method		

Common Name		
Latin Name		
Origin		
Signals		
Soil Preference		
Benefits/Uses		
Appearance		
Control Method		

Common Name	
Latin Name	
Origin	
Signals	
Soil Preference	
Benefits/Uses	
Appearance	
Control Method	

Common Name	
Latin Name	
Origin	
Signals	
Soil Preference	
Benefits/Uses	
Appearance	
Control Method	

Common Name	
Latin Name	
Origin	
Signals	
Soil Preference	
Benefits/Uses	
Appearance	
Control Method	

Common Name	
Latin Name	
Origin	
Signals	
Soil Preference	
Benefits/Uses	
Appearance	
Control Method	

Common Name	
Latin Name	
Origin	
Signals	
Soil Preference	
Benefits/Uses	
Appearance	
Control Method	

Common Name	
Latin Name	
Origin	
Signals	
Soil Preference	
Benefits/Uses	
Appearance	
Control Method	

Melanie Holsti is an avid tea lover with a deep appreciation for the simple joys of life. Whether she's sipping a cup of Scottish Breakfast tea, capturing the perfect shot with her camera, or getting her hands dirty in the garden, she embraces each moment with enthusiasm.

As an author and small business owner, she pours her focus and creativity into everything she does.

Her background is as varied as her interests—she's worked in hospitality, retail, livestock farming, writing, and, of course, gardening. Over the years, she has experienced the triumphs and challenges that come with growing food and homesteading in diverse climates, from the harsh prairie of South Dakota to the woods and mountains of Missouri and Tennessee.

In 2018, Melanie and her family took on an exciting project—restoring a charming 1917 Craftsman bungalow. With a lot of hard work and a shared vision, they transformed it into the Ironstone Cottage Tea Room, where guests could gather and enjoy a hot cup of tea or lunch with friends in a cozy, welcoming atmosphere.

These days, Melanie, her husband, and their daughters reside on a small farm in Tennessee, living in a beautiful Federal-style home they built together. Their homestead is shared with an assortment of animals, including a peculiar flock of barn cats, free-ranging chickens, and noisy ducks.

When she's not tending to the garden or caring for her animals, Melanie finds joy in writing books, sampling new and exotic teas, and cheering on her favorite professional hockey team. No matter the season, she's always pursuing her interests and creating a life filled with warmth and, of course, plenty of tea.

www.ironstone-tennessee.com

IronstoneTennessee

IronstoneTennessee

amz

amazon.com/author/ironstonetennessee

ironstonetennessee

secret resources web page
Password: IronstoneBooks

Your feedback and reviews are priceless. If you have suggestions for improving my gardening workbooks, I'd love to hear them. If you have a moment, please consider leaving a review on Amazon or my website.

Other Titles by Melanie Holsti:

The Fruit, Flower, Herb & Vegetable Gardener's Journal

The Fruit, Flower, Herb & Vegetable Gardener's Planner

The Ironstone Garden Journal

The Ironstone Garden Planner

The Ornamental Files

The Houseplant Files

The Herb Files

The Orchard Files

The Vegetable Files

The Ironstone Forager's Notebook

Journaling Your Weather

Journaling Your Health (A Collection of Women's Wellness Journals)

Ironstone Prayer Journals

Intercessory Prayer Journals

Tea & Tisane Blending Workbook: A Tea Mixologist's Journal (with Jennifer C. Petersen)

Made in United States
North Haven, CT
08 November 2026

21928348R00104